The Real Universe.

I0489224

The Real Universe:
The Real Physics of the Universe, Time, and the Matter of Space.

The introduction of the founding of Pouliotion Physics by Pouliotion Cosmography.

Pouliot's Theory of Absolute Specificity.

A new Hypothesis of the G.U.T.

By

Russ Pouliot.

The Real Universe.

The Real Universe.

How to use your new Pouliotion Physics Manual!

The best way to read this hypothesis is from beginning to end. Elements herein are and will be controversial. This paper is of bottom up construction. Later proposals are supported by initial theories.

The Real Universe.

Forward.

What is this and why did I write it?

This is a hypothesis of the physical nature of the universe. How the atom is constructed and how this particular universe we live in began, is constructed and works.

This is proposed to be the real basic foundation of what is now understood to be the: Grand Unifying Theory or G.U.T. I say basic foundation because 'Pouliotion Physics' (as I have named it) will only open the door to the true understanding of matter and/or physical existence. Pouliotion Cosmography (the fundamental principles that define 'Pouliotion Physics') will be taken a long way from its introduction here.

This hypothesis called, 'Pouliotion Cosmography' which leads to Pouliotion Physics is more of a founding. It is inspiration and deduction, based on empirical data and information that already exists. It stands on the shoulders of all who have wondered after the most basic understanding of our physical existence. There have been many.

Explanations and some predictions here will be verified by past observations and new observations based on the new understandings provided by Pouliotion Physics.

I have also wondered after the true nature of physicality. I have stated about myself and I always will say of myself that I am a very spiritual man. I have provided this hypothesis because I have to. I was told to. It was given to me to give to this particular Earth. I'll let it go at that.

The Real Universe.

Introduction.

More amazing to me than nothing, is something.

I wouldn't suppose there is any person that has lived who has not wondered about the true nature of the universe. From what I understand from the history of the world's collective consciousness, such knowledge was/is believed to be unattainable. That is, most people that have ever lived believe understanding how eternity is created is not really knowable.

The mind is the ultimate microscope, or telescope. Observations with associations, deductions, and inductions by the mind are the ultimate learning and discovering tools. I maintain that there is nothing that cannot be observed, discovered and learned.

When I was young, child friends and I would hang around at the Y (YMCA). They gave Judo and Karate lessons there. More than once I, with a group of neighborhood kids would approach the black belt teachers as they mulled with each other while the next class was gathering. Our ongoing single question, sometimes asked by me and other times asked by one of my fellow ragamuffins was, "What's the 'main thing' of Karate? or Judo?)."
"Karate is mainly boxing." It was a good answer.
"Judo is mainly wrestling." Another good answer.
Somehow, our child's' minds had a faith and a belief that we could be taught, with the answer to one question, a martial art and go out and beat up any kid in the neighborhood. This without having to put in years of study and dedicated work.

I eventually achieved a black belt in judo. I understood the answer to my childhood question (after years of study and dedicated work). Judo was the art of using what is called a, "Moment of Inertia" against an opponent. Layman's terms would be, "Using a persons weight against them."

The Real Universe.

Other than inertia, judo has little or nothing to do with the real universe. The moral of my childhood question had everything to do with the real universe.

My 'universal' question derived by the child's mind of faith applied to the grown up world.

"What's the main thing of the universe?"

However it happened, I knew from a young age that I was allowed to ask that question and had a faith that the answer could be known. My years of study and dedicated work on finding the 'main thing' of the universe where inherent in me. Anyone who knows me, knows of my bulldogedness. When I get attached to something, a question or whatever, I won't let go. Over years I paid attention to the state of the science of cosmology. What was known from the macro (Astronomy) to the micro (Particle physics, atom smashers, chemistry). Also the pioneers, Hubble, Einstein, Heisenberg. All of this sunk in. It was inside cooking, bubbling, refining.

From studying judo (and all else) life went on for me, evidently the 'main thing' was always on my mind. It was like my wife says, "You're always thinking anyway, even in your sleep." And I do, and I was. For the cosmographer me, the child (me) was the father of the man (also me) and I was finding the answer all the while.

I never seriously considered becoming a particle physicist or a nuclear physicist or a quantum physicist. The 'main thing' was always on my mind, though.

Call it serendipity; call it inspiration. Don't call it luck or some result of randomnicity, but recently it dawned on me what the answer to my universal question was. Postulations herein are based on what was realized as the result of my somnambulations and conscious thinking about the 'main thing' of the universe over a lifetime.

Most basically, the Main Thing of the universe is Space.

The Real Universe.

PC (Pouliotion Cosmography) will undo current labyrinthal thought that space is a void, and introduce the knowledge that perceived nothingness is really the primal and premier something of existence.

With PC, (Pouliotion Cosmography) broader understanding of space and how space creates the universe and the universe is created out of space will be set forth. The distress this puts on currently held beliefs about presently held cosmology will be worked out.

At the heart of PC is understanding that space is not nothing. It is said that space bends and is more or less dense, yet there is some vague assumption that 'it' is nothing, a void. How can nothing bend? Space is something, a fabric. Space consists of particles, near but not infinitesimal. Space particles would be a good model of the infinitesimal or a point.

Currently accepted beliefs about the 'states of matter' will be expanded and the concept that there are really four states of matter will be revealed, developed and seen as self-evident. This new knowledge will support and explain itself.

The spatial particle will be described and defined. As its nature is brought to light, the relationships explaining physical phenomenon of such things as gravity, inertia, the speed of light, etc. will brought to light and be more fully understood and explained.

It will be seen again as it has many times in science how the micro dictates the behavior and existence of the macro. How we can deduce the micro from observations and measurements of the macro.

It is time now, for this world to have a 'Quantum jump' if you will pardon the intentional pun. A quantum jump in knowledge of the cosmography of our universe. I differentiate between cosmoLOGY and cosmoGRAPHY. Cosmology being the myths and/or legends of how the universe is created and runs. Cosmography being the facts of the same.

The Real Universe.

This new understanding will be recognized by the many branches of science. It will fill in gaps and fulfill many areas of science where the physics, chemistry etc. hits the,
'we don't know yet' wall.

This new theory could be likened to medical science before bacteria and viruses where known to exist.

More amazing to me than nothing, is still something. So in space, I found a lot of really amazing something that was supposed to be nothing. The microscope/telescope of the mind has prevailed again, and with the answer to one question, we go on now and take another step out into our universe. Years of study and dedicated work lay ahead of us. I'll stick my neck out here and say it will be an eternal labor of love for mankind.

RP.

The Real Universe.

1. The way we where.

The following is what the state of the universe seemed to be before PC:

The universe, all the matter in the universe seemed to be compressed down to the size of a neutrino or smaller. Somehow it expanded or exploded and has been doing so ever since. How that could ever be?

Nuclear structure.
Atoms are like micro planets? Electrons are orbiting the atom. Smashing atoms to find out what atoms are like seems to be saying that atoms are the pieces that they are smashed into. A bunch of broken up pieces does not seem ever to be able to tell how the intact thing was made or its state of existence.

Gravity. Seemingly there is a force inside all matter that pulls on other matter. There is some phenomenon that reaches out and pulls us to, supposedly the center of the earth. There seems to be some force between planets in that they reach out and pull on each other. Same with our sun, stars. Not understandable really, another, "We don't know yet."?

Inertia. Again, how? Why? As sparsely understood as gravity. Newton described the behavior quite well, but how and why? No clue.

Missing matter. Dark Matter?!
So, how can: " - - - -96 percent of the universe be made up of stuff astronomers can't see, detect or even comprehend - - -."? (Moskowitz, Space.com, 5-12-11)

Light. Light travels through a void? Why does it 'travel' up to a certain speed and no more? Light can bend, does that mean that space can bend? How can space bend if it is nothing? How could light or anything travel through nothing? If there is no medium to travel through, how could anything travel through it?

Time. Yeah, right.

The Real Universe.

There are good observations of the universe and some misses.

This hypothesis has the intended purpose of clearing up understandings of these observations and phenomenon and many more. A lot of what is stated above as current knowledge of the universe is from the13th century. All of the above will be explained with PC, it is explainable and is explained.

The Real Universe.
2. Space and/or The Universe: Start/Stop.

The hypothesis begins with the non-reality of nothing. Nowhere is there a void. Nature abhors a vacuum and does not allow a void.

This postulation defines that the primal universe eternally begins and ends with an infinite number of Space Particles, they already exist, they have always existed, there has never been a state of existence in eternity in which they did not exist. PC defines the universe in this specific way. In PC, the universe is defined as an endless number of contiguous, "Space Particles" along three axis.

The Real Universe.

3. Space Particles. The Particulate.

A single space particle is termed, "The Particulate," it has a definite structure. The Particulate is made/constructed out of Space/Spatial Material/Space Matter.

"Spatial Material/Space Matter/Spatial Fabric" is a 'given' in this hypothesis. It is <u>the</u> basic element of the universe. It is the newly revealed element; the knowledge of it originates here as postulated in PC.

(Figure 3.)

The Particulate is called out as a cube within a cube. It is constructed of 'spatial material'. It has mass, relatively small mass, but it's there. The 'filaments' are made from denser spatial material, the nucleus is also composed of denser Spatial material. The area inside the particulate that is not filamental consists of a thinner/less dense spatial material, also founded here in PC.

Because of the thinner/denser nature between the filaments and non-filamental area inside the square area of a cube/particulate, there is a potential difference, hi/lo pressure difference if you will, between the filaments and non-filament areas inside any given cubic area of any given Particulate. This will be seen to cause directionality along any given length of a filamental segment. (Directional push/pull vector)

The Particulate has a bias(Figure 3.1), that is, the outer and inner cubes are not normal angularity to each other (with the exception of Primal Space). The degree of angularity is highly dynamic depending on the amount of energy stress load on the particulate at any given point in 'time'.

The non-normal angularity can be seen in figures (3b, 3c) and are termed, 'a bias', or 'the bias'. The bias angle is a critical feature of the

The Real Universe.

particulate. PC postulates that filament directionality and the bias angle of the particulate account for almost all the physical behavior in the universe; gravity, inertia, light speed and pin wheeling galaxies.

How this was determined will be self-explanatory as the postulation continues.

There are three other states of Space Matter: solid, liquid and gas, these are known herein as 'Hard Matter'. Hard Matter (solid, liquid, gas) is derived from Spatial Material/Spatial Matter. So, in PC there is one element (Spatial Matter), three derivatives of the one element (Hard Matter. Solid, liquid, gas.), and four states of matter.

The Real Universe.

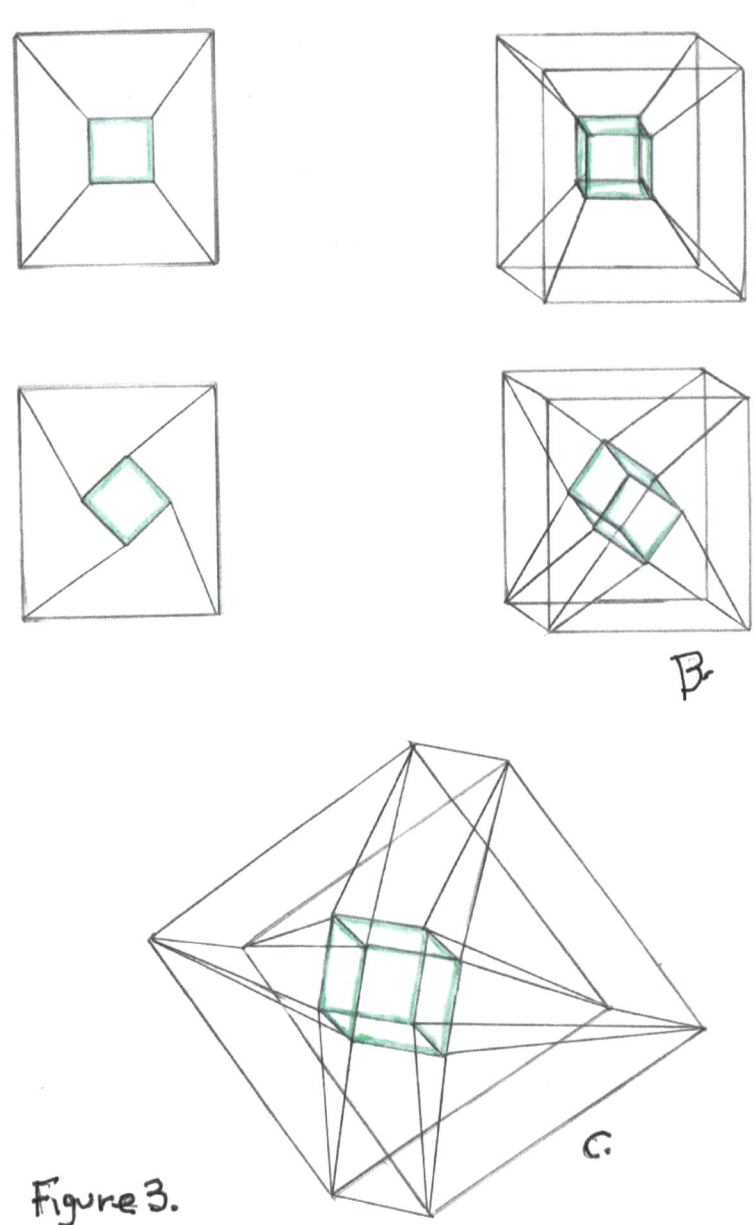

B.

C.

Figure 3.

14

The Real Universe.

The Particulate.

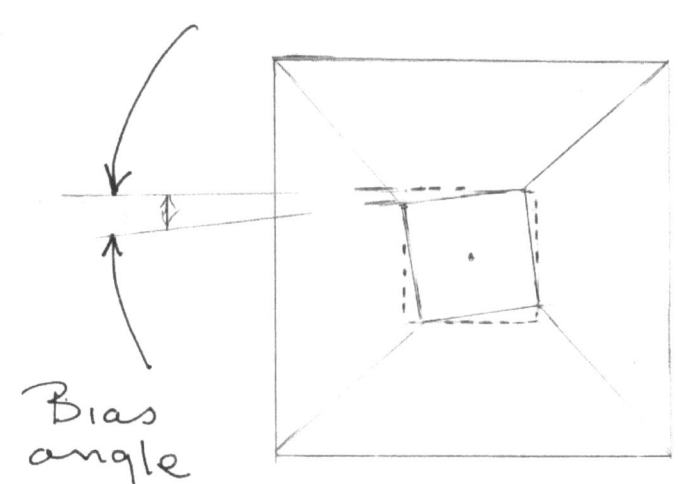

Bias angle

Figure 3.1

The Real Universe.

4. The Primal Universe.

Before there was anything even a Big Bang there was still not nothing. The primal universe consisted of space particles. (Figure 4.) As illustrated, the universe was a homogeneous infinity of contiguous 'Particulates' endlessly along all three axes. This might be compared to endless child building blocks being stacked together.

In this state there is no bias angle, the Particulate cubes all have normal angles. This postulation assumes there is no Hard Matter and no time. There is plenty of Space though, Space particulates.

The Real Universe.

Primal Universe

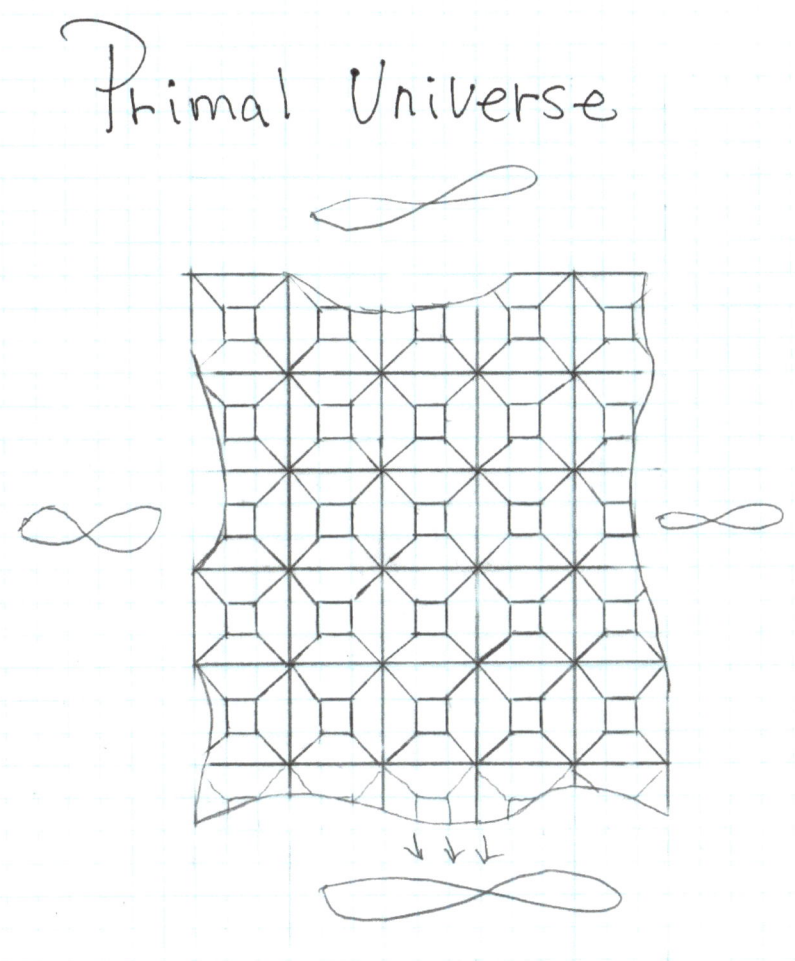

Figure 4.

The Real Universe.

5. Post Energy Event Universe.

Space and the space Particulate/s are different before an energy event than after. (Figure 5.) 'Energy Event' is a 'Big-Bang'. Starting with a Big Bang, after, and in the vicinity of a Big Bang, Space Particles are now under a stress. Because of impacting of a large amount of energy the Particulates become 'charged' or as was said 'under stress' or even 'under tension.'

This causes the 'bias angle' to 'torque', and the cube angles to not be normal any longer, or forever after. This might be compared to the seeds in a pomegranate fruit. (Figure 5.B)

The Real Universe.

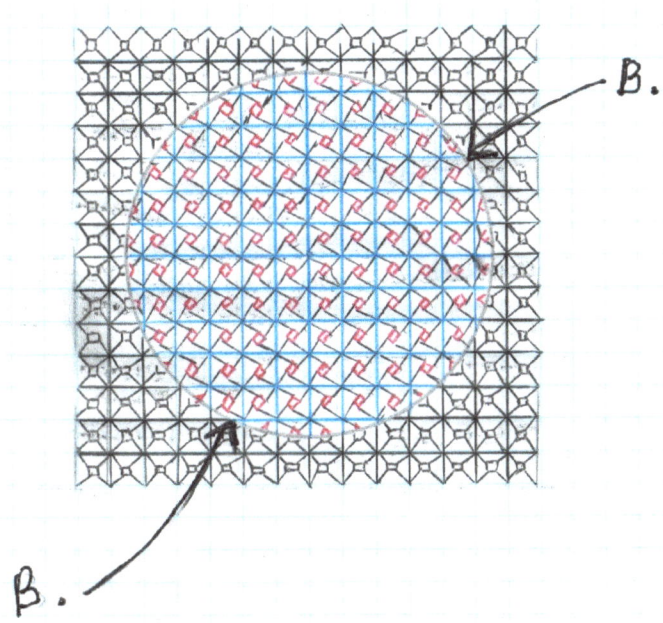

Figure 5.

The Real Universe.

6. CREATION OF HARD MATTER: Hard Matter Creation Replication Cascade. HMCRC.

Energy event.

It is a given here that a 'big bang' occurs at point of origin (0,0,0) and that there is no matter at all extant. The energy event itself is a given. Energy is released in and of itself and is not the result of all the 'matter' in the universe being so compacted that it explodes. In the PC Big Bang there is only an infinite lattice of contiguous space particles before the energy event. There is no matter of any kind. (Figure 4)

In the locality of an energy event, (This energy scenario is a Big Bang) contiguous Spatial Particulates proximal to the energy event (Figure 6.A) are fused into Particulates distally contiguous to them and to the energy event (Figure 6.B). This happens in an expanding cube as the explosion accelerates.

What is happening with this energy event is that spatial particles closer to the energy event (6.A-D) are compacted into spatial particles that are contiguous to them distal to the event (6.B). This causes a volume, theorized to be a very large volume of 'space' to be compressed/welded/fused in a specific manner into a small number of adjacent Particulates to become what is now understood to be 'matter', or a basic building block of hard matter. The Spatial particulate in this event would be the 'mold' into which other spatial particulates have been compressed. Like anything, which has been pressed into a mold, the opposite image is created. So we have the very basic hard matter now 'materialized', 'synthesized' in the reciprocal/mirror structure of The Particulate.

The newly created/fused hard matter particles are compacted again into spatial particles again on the distal side of the event creating denser structure of hard matter. This continues until a reverse critical mass is reached and there is not enough energy for additional compaction or fusing to occur and the first meta-stable state of 'hard matter' is reached. This structure is theorized to be the proton or neutron (Figure 6.C)

The Real Universe.

depending on how many fusions/synthesis cycles (odd or even) it went through to reach the first stable reverse meta-stable state and become true hard matter.

The 'hard matter' being created is cubic and continues its geometric expansion. As more hard matter is created, it is created in the form of cubes. The cubic shape of the hard matter of the entire big bang increases exponentially (6.D), building larger and larger cube that is the identical but reciprocal construction of the original neutron/proton cubes. A cube of hard matter of hydrogen is now forming in the shape of the construction of the neutron/proton and it is the size of the universe. The hard matter is now moving and begins spiraling because the bias angles of the hard matter are vectoring/shearing in ribbons with the bias angle of the spatial matter they are moving through; essentially the hard matter (hydrogen) is pin wheeling through the fabric of space.

In a small amount of time there is an object of hard matter (hydrogen) (6.D) billions of light years across. This object is the same-mirrored reciprocal shape construction as the cubic particulate. The universe size hard matter object has a cube within a cube construction, the filaments being light years in diameter. As has already been stated, the hard matter is moving and spirals through space like a pinwheel, because space is biased as well as the nucleus of the hard matter.

As the spatial particles are synthesized/fused in an upward chain, the hard matter in the reciprocal/mirror image of the cubical shape of the spatial particle is maintained, so the building blocks of hard matter are being replicated and remain a cubical within a cube as the energy event continues to expand.

Another given in PC is that Spatial Particulates are not moving. Space Particulates cannot go anywhere do to the fact that they are all next to each other. 'Space' then has the ultimate and infinite moment of inertia.

The Real Universe.

It is the ultimate, non-moving fabric that is truly universal. Since space particulates have the ultimate moment of inertia, an infinite XYZ lattice of spatial particulates really are the only stationary things in the universe. Like any fabric, Spatial Particulates can be disrupted stretched, compacted and are malleable and elastic. A high-energy event begins fusing, explosion welding the spatial particles together.

This compacting/fusing of Particulates into Particulates creates the most basic hard matter; it is part of the 'Nucleosynthesis' Hoyle spoke of.

So, it is postulated that Hard Matter 'materializes' from the fusing/crushing of the much finer 'Spatial Matter or Fabric' into itself. PC states then that: The basic building block of all 'Hard Matter' is hydrogen and the building block of the neutron/proton IN the hydrogen atom is the Spatial Particulate. 'Hard Matter' is at present thought to be 'All the matter there is in our local universe'.

Hydrogen is observed to be 74% of all 'Matter' really, 'Hard Matter' in our universe. Hydrogen is the most basic and plentiful of the, 'Hard Matter Elements' to be observed in the universe.

A big bang creates vast amounts of 'hard matter' (mainly hydrogen) 6.2. and disperses it spiraling throughout the universe it is creating. A big bang creates what we understand to be a, 'Light Horizon Bubble.' That is all we can observe with present technology.

At the event of a big bang, hard matter is created and exists for the first time, and time begins. Time, basically is the movement of one thing through or past another. In Pouliotion Physics, time is the movement of hard matter (liquid, solid, gas) through stationary Spatial Fabric. Passing through space particles has potent effect on hard matter, namely, [Gravity, Inertia, Time.]

The Real Universe.

The Real Universe.

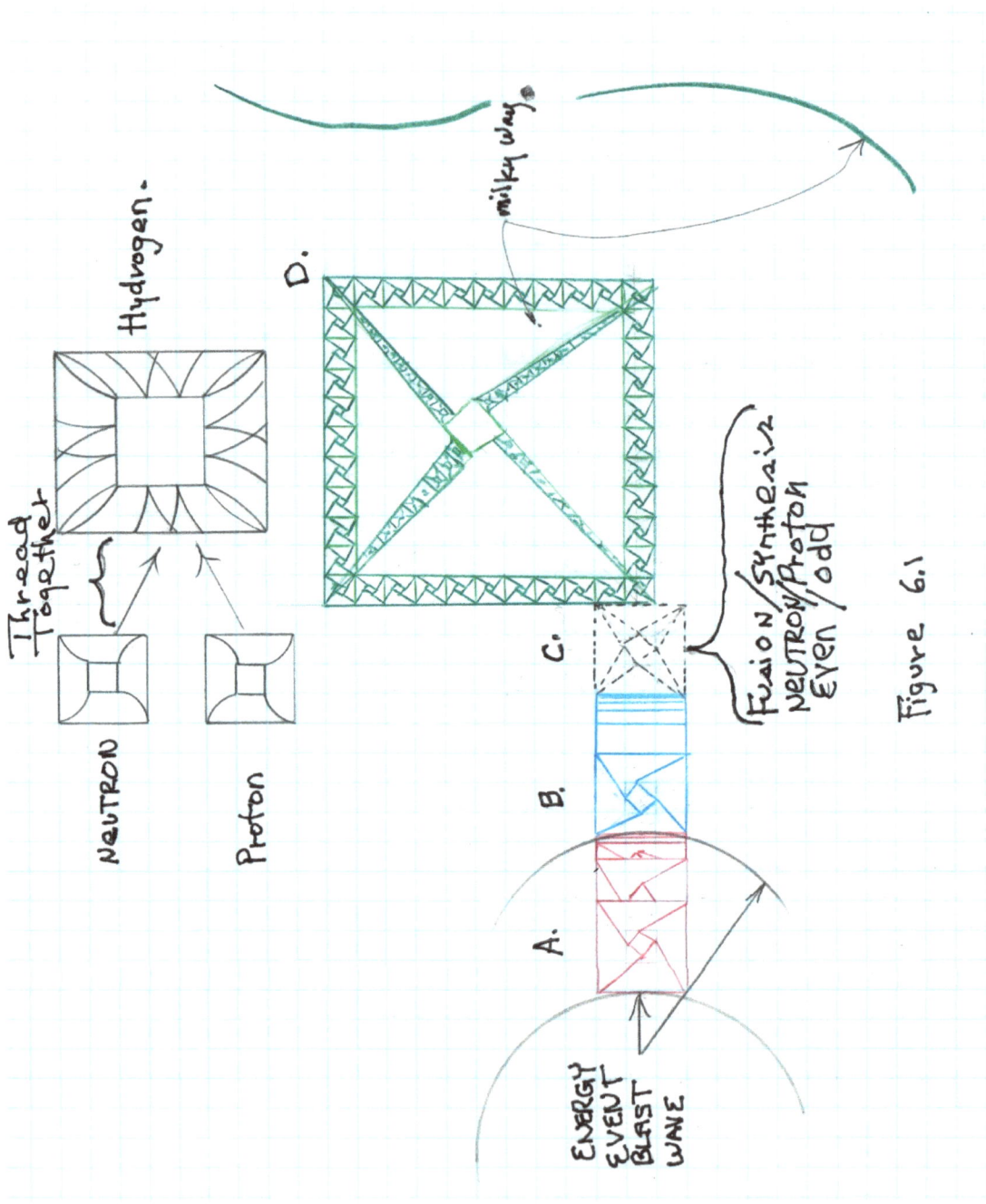

Figure 6.1

The Real Universe.

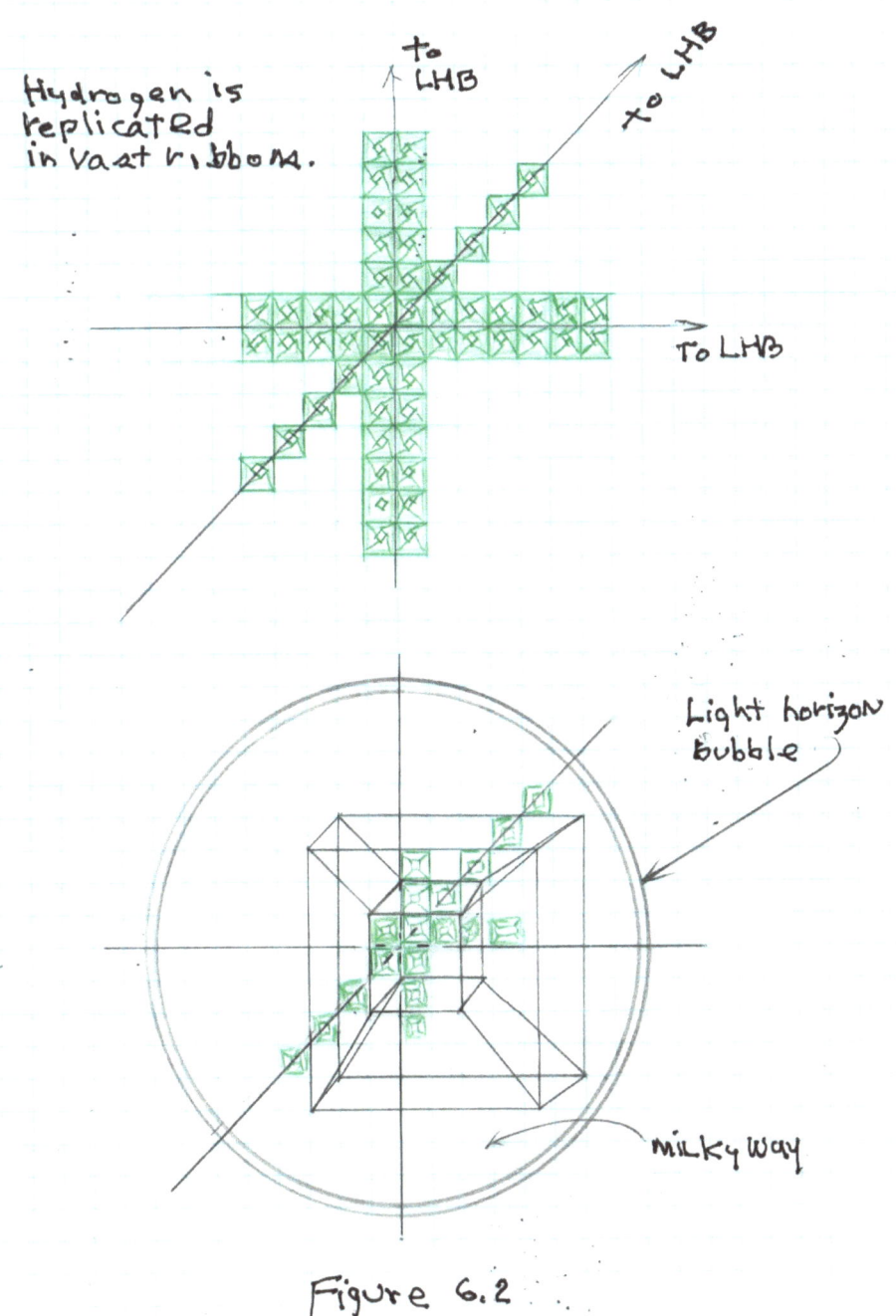

Hydrogen is replicated in vast ribbons.

to LHB

to LHB

To LHB

Light horizon bubble

milky way

Figure 6.2

The Real Universe.

7. Gravity.

One of the main effects of Hard Matter creation (HMRCC) is gravity. The algorithm that explains gravitation is as follows:
Gravitation is analogous to buoyancy in water. Take an ice cube floating in water in a large glass jar (Figure 7.1). When another ice cube is added to the water in this jar, the original ice cube momentarily rises with buoyancy.

The analogy is this: The jar is our Light Horizon Bubble. The water is the Spatial Fabric/Fluid in the bubble. The ice is Hard Matter. Hard matter creation is another ice cube being added to the water. If one where to be standing on the ice cube in the water when the second ice cube was added, the person standing would feel the rise or buoyant rise. If more ice where to be added to the water at a rate of $32ft/sec^2$ the person standing on the ice would feel gravity. This is Einstein's elevator ascending at $32ft/sec^2$. Note the 'squared' nature of gravity. The Particulate is square.

The ice cube in the jar of water serves well enough to demonstrate the basic property of gravity as buoyancy (Space buoyancy). With gravity, the fluid (space, which is not nothing) is all around us and any hard matter object. It is seen then that gravity is a mass/space buoyancy or fluid pressure that pushes us toward the center of a hard matter object from an infinite number of points around us at all distances from us (7.2). We have Pascal to thank for the understanding that pressure transmission through a fluid is instantly equal at any physical distance. Thus gravity is a buoyant pressure of spatial fluid exerted along the entire surface area of any hard matter object. On a para spherical object such as planets or stars this appears as if there is something pulling an object on the surface toward the center. It is buoyant space fluid pressure from infinite points around the object necessarily pushing the object to the center that causes this result though. Similarly, presently believed gravitation between planets, stars and all hard matter objects are really the result of spatial pressure from the very wall of our Light Horizon Bubble.

The Real Universe.

The mass creation rate of 32ft/sec^2 was mentioned. This is the particular calibration for us here on earth. It is obvious that if the blocks of ice in the analogy where bigger or smaller, or had more mass or less mass. The buoyancy would be at a different acceleration rate; in other words, different rates or levels of gravitation. This is confirmed in the observations of the gravitation on other larger or smaller, more dense or less dense planets and also the gravitation of different stars.

Also: The pushing of more hard matter into the fluid of space causes a wave build up around the surface of hard matter objects, causing a 'bulge' of gravity around planets and stars. The Spatial Fluid is a little denser around the surface of planets and stars, this causes Einstein's' effect of the starlight bending around the sun.

Hard matter creation resulting from a big bang goes on for a vast amount of time. It does not however, go on indefinitely. A big bang is the start a more powerful and nearly perpetual hard matter-creating engine, The Super Nova. For super novas to be created, stars need to be created first. Star formation in this hypothesis is now more clearly understood in the light of hydrogen creation and deeper understanding of gravity.

The Real Universe.

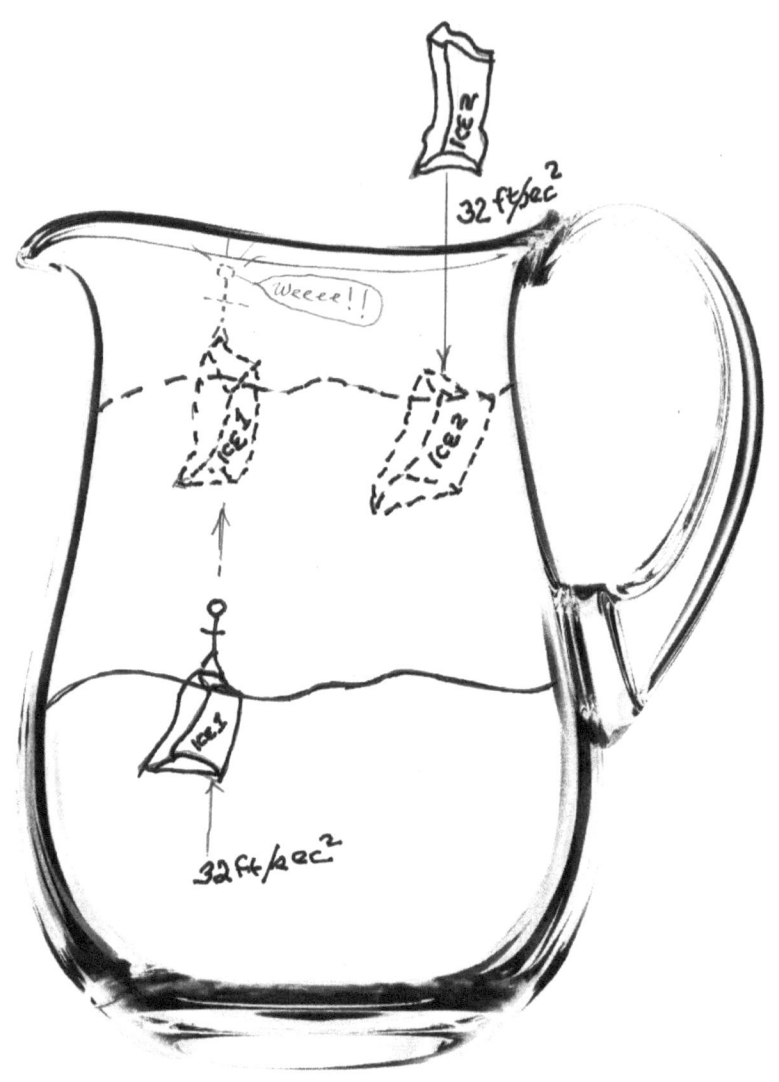

Figure 7.1

The Real Universe.

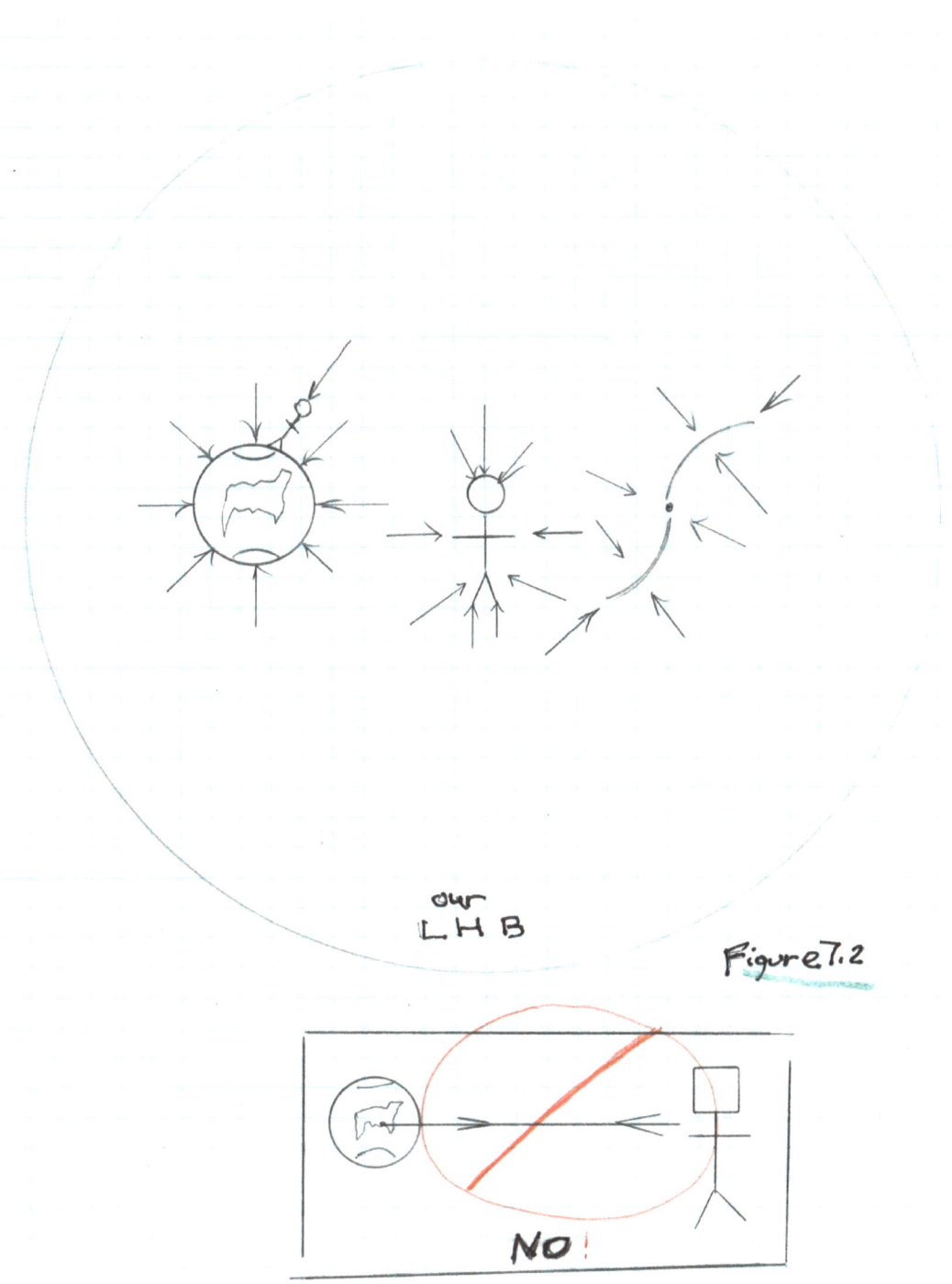

our
L H B

Figure 7.2

NO !

The Real Universe.
8. PC's postulation of Star Formation. (And planet formation)

Star formation can now be better understood with a deeper understanding of gravitation thru PC. Stars are currently believed to form when large clouds of hydrogen 'collapse' this due to their gravitation 'pulling from within.' With PC's more accurate and in deeper understanding of the phenomena of gravity it can be seen that clouds of hydrogen are 'pushed/compressed' together from a universal number of points all around them. That is why hydrogen atoms would be pushed together even from a state where they are not dense enough, by current understanding to 'be attracted to, or pull on each other.' There is no need for a: " - - - kick from a nearby supernova explosion, collision with another gas cloud, or the pressure wave of a galaxy's spiral arms passing through the region - - -" to cause a cloud of hydrogen to 'collapse' or 'fall in on itself', which it does not do anyway. Clouds of hydrogen are 'compressed' by gravitation to form stars. (Figure 8.1)

Initially the gravitation (i.e. fusion of Spatial Particles to Hard Matter or hydrogen causing gravity, 'putting more ice in the water') to form the first stars is supplied by a big bang. The hydrogen created in the big bang spirals off and is compressed by the big bang's initial gravitation to form the first stars. At some point older and/or larger stars use all or near all of the hydrogen they are fusing in their cores to keep an outward pressure in the core equal to the gravity weight of the star shell. This reduces the hard matter density of the core relative to the density of the shell immediately around or encasing the core. When the fuel keeping the core pressure up runs out, or runs low enough the denser hard matter around what was the nucleus of the star is compressed in or collapses.

The collapsing core of a super nova in the fluid of space is the ultimate depth charge. The resulting super nova explosion (or really implosion) creates an energy event large enough to again fuse Spatial Matter/Fluid into neutrons/protons and therefore hydrogen. Seeing super novas in stop motion, how's that

The Real Universe.

for a moment of inertia? Also PC is in concurrence with presently held cosmological beliefs that hard matter of heavier and heavier elemental weights are created/synthesized and dispersed in a super nova implosion/explosion.

Micro structured is mirrored/replicated upward, like crystalline formation to any macro size at any 'time' in the progression of synthesis.
Look at super novas. The Spatial Particulate filaments are still visible at the galactic size level. (Figure 20)

In the PC supernova scenario there is an interesting premise for a treatise on isotopes.

The immense difference between the density of Space Particulates (space) and the least dense Hard Matter (hydrogen) is seen here to be the potential difference that perpetually fuels energy transference in our light horizon bubble. Truly, Conservation Of Matter is maintained in that no energy is lost. This process is the ultimate heat pump/exchanger that perpetually powers the universe.

The Pouliotion Universe begins with a primal energy event or big bang. It is a 'clean' big bang in that only an energy event is present; there is no pre-existing matter present, compacted or otherwise. The energy event creates Hard Matter out of Spatial Fluid by welding or fusing large amounts of Spatial Matter or Fluid. The compressed or welded results are protons and neutrons that combine to make hydrogen. Hydrogen is propelled outward in spiraling arms in vast clouds. Hydrogen clouds are compressed by universally present gravitation to form stars. Stars go through their cycles to nova, being engines for more nucleosynthesis and fusion furnaces for the heavier elements.

When novas get underway the universe has reached a steady state of self-sustaining matter creation. This is the 'Steady State' existence that Hoyle spoke of. Novas are readily available for observation. PC concurs that there is a high probability that there is one super nova per second in our LHB (Light Horizon Bubble). Hard Matter creation is

The Real Universe.

now being done mainly by super novas creating an avalanche of basic hard matter and gravity on a constant, eternal basis.

A light horizon bubble that has one super-nova per second going on inside it does not seem to be much of a candidate for any kind of contraction. A steady state light horizon bubble seems to be more the case. At this point the concept of entropy is put out and put to rest.

The Real Universe.

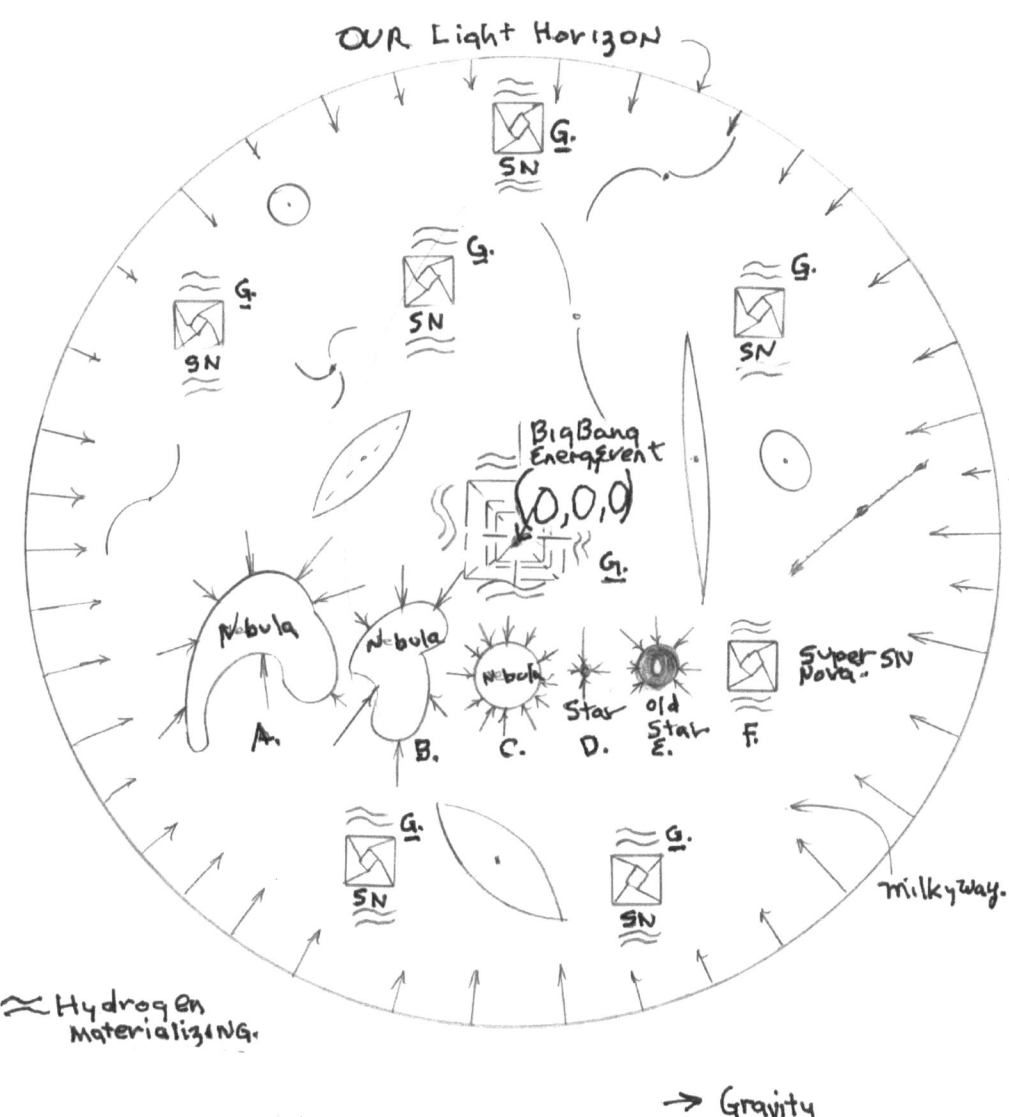

Figure 8.1

The Real Universe.

9. Maybe a segue here? Odd synth even synth, red shift blue shift.

In a steady state universe or light horizon bubble, PC would postulate that red shift or blue shift galaxies would either be clockwise or counterclockwise rotations. This because of the bias of the particulate spatial material they are traveling through, and the bias angle of whatever hard matter that is moving through space, i.e. the Odd/Even number of nucleosynthetic cycles in formation of proton/neutron. It would seem to be a confirmation of the particulate bias premise itself if red or blue shift would correlate to clock or counter clock rotation, i.e. left or right-handedness.

Also the images produced by accelerators show spirals. Pieces of something curving away or being deflected from a straight line away from the impact point. This would also be attributed to a bias angle in the space particulates that the results produced in the accelerator are traveling through. (Figure 9.1)

The effects of something traveling through space and traveling and being deflected by the bias angle of Spatial Particles is readily seen when two mirrors are faced to each other with someone in between. One always wants to tilt their head to see around the corner. This seeing images curve out of view, it is proposed is a good way to see the effects of light being deflected from straight travel because the nucleus of The Particulates (space) the light is traveling through are at a slight angle to the incidental light. The light bouncing back and forth between mirrors accentuates this slight bias angle in the space where the light is traveling between the mirrors, making it look like the image curves around infinitely.

The Real Universe.

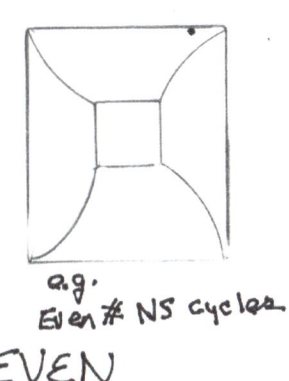

a.g.
Even # NS cycles

EVEN

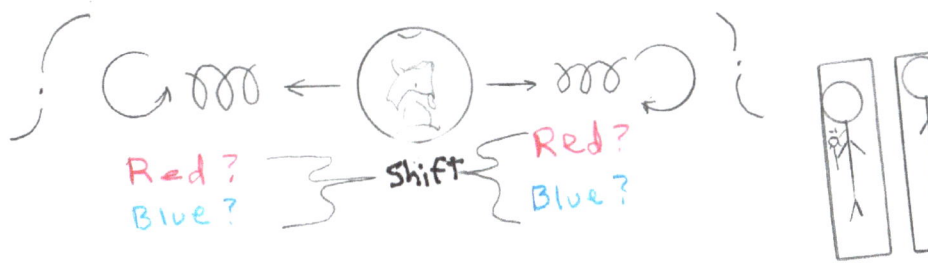

Red?
Blue?

Shift

Red?
Blue?

Propeller
impeller?

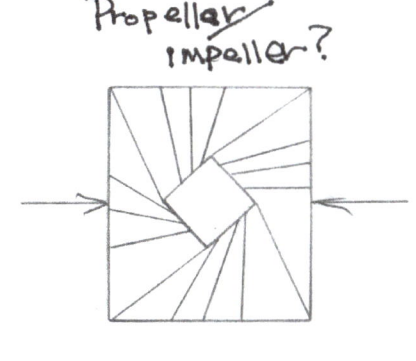

e.g.
ODD # NS cycles

ODD

Figure 9.1

The Real Universe.

10. Black holes.

A steady state universe would not be possible without black holes. Black holes are cosmic or universal capacitors. They keep hard matter creation at a smooth and even rate giving us here on earth a constant $32ft/sec^2$ of gravity. Black holes compress hard matter enough to squeeze spatial material/particulates out (Figure 10). Recycling spatial material is another good premise for a treatise on isotopes. 'Space', or 'Spatial Particulates', or 'Spatial Particles' jet out of black holes because they (Spatial Particles) are not affected by gravity, just as the water in the glass jar is not in and of itself affected by another ice cube placed into the water. A black hole has enough 'gravity' to cause or allow the hard matter falling into it to be reverse fused or 'unsprung', releasing the spatial particles, which are then vented or jetted out into 'space' again.

Black holes complete the nucleosynthetic cycle of --> Spatial Fluid/Particulate to --> Hard Matter back to --> Spatial Fluid/Particulate. Mass and energy are conserved. Our Light Horizon Bubble, it is postulated here is truly a perpetual motion machine. Black holes are the finishing step in the conservation of matter and (it is postulated here) making energy transmission and conversion of the big bang and of super novas eternal in our light horizon bubble.

The Real Universe.

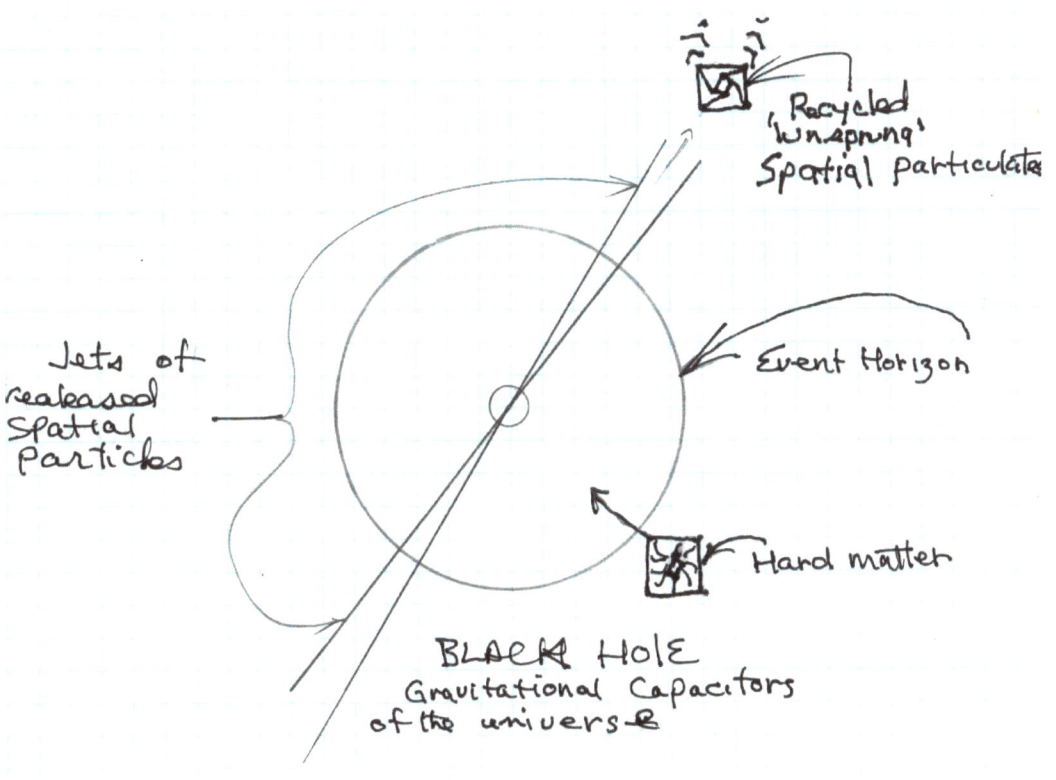

Recycled 'unsprung' Spatial particulate

Event Horizon

Jets of released Spatial Particles

Hard matter

BLACK HOLE
Gravitational Capacitors
of the universe

Figure 10

The Real Universe.

11. Dark Matter.

Spatial Fabric has a mass. It is postulated here that the 96% of matter that is termed, 'dark matter' is the Spatial Material that constitutes 'the universe', the remaining matter of the universe that's not hard matter. Spatial Matter is transparent and very fine. It is not detectable with present technology. Its effects are readily discernable and measureable. It should be possible to crunch the numbers to find the square area and mass of a single Space Particle. The known universe being what?: a bubble 93 billion light years in diameter. The 4% of mass (hard matter) is basically known. So, 96% of the calculated mass of the universe remaining divided by what's remaining of the volume of a 93 billion light year diameter bubble.

The Real Universe.

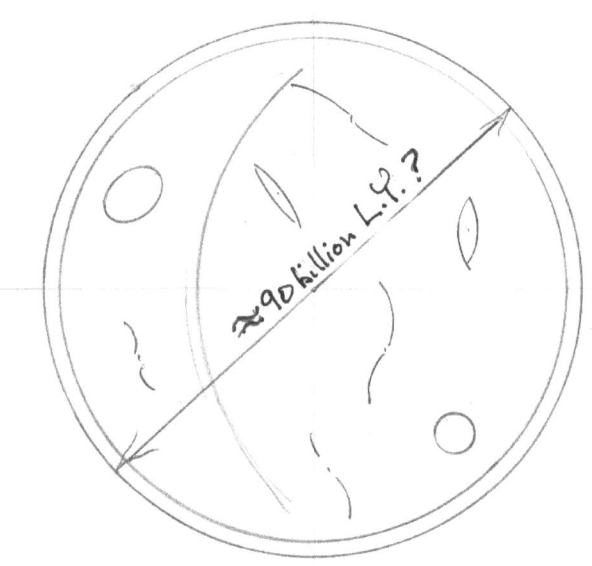

Figure 11

The Real Universe.

12. Inertia.

PC has already defined pre-big bang and post-big bang space. Post-big bang space will be used in this quite good postulation of inertia. The Spatial Particulate is reiterated here as endless cubic units of contiguous fabric. (Figure 12.1)

Again the 'filaments' are composed of 'more dense' spatial matter or fluid, the non-filamental areas of the Space Particulate are composed of 'less dense' spatial fluid. Because of the density difference and non-normality of the outer and inner cubes there is a 'directionality' of the filaments and moreover the whole Spatial Particulate cube within a cube. (Figure 12.1) Each Spatial Particle has a 'pull' if you will along its filaments toward its own center or nucleus. (Figure 12.1)

A hard matter object entering an area of space influenced by one or any number of Spatial Particulates acting as one is, 'pulled in' or 'falls in' a little to the center or core of the virtual particulate of space. The same object, when it travels through the core or center and out to the opposite side of the particulate, is 'held back' a little or has to 'go uphill' a little.

An object moving through space is, 'held' in its velocity of travel by innumerable 'pullings' in opposite directions, with and against the direction of travel of the object. To increase the velocity of the object, innumerable little 'braking tugs' need to be overcome with energy; to decrease the velocity, innumerable little 'pulling' or 'accelerating' tugs need to be overcome, again with more energy.

This is the PC theory and postulation of inertia. Known phenomenon considered at this point are, "Starting Friction and Moving Friction." PC inertia explains why starting friction is greater than moving friction.

Also, a spinning top. It can be seen that all the little 'pullings' and 'draggings' are what keeps the top up as well as explaining gyroscopes.

The Real Universe.

The phenomenon of increased inertia with increased mass correlates to more 'Spatial Particulates' interacting the increased hard mass, or increased number of atoms of the object, adding more 'little pullings' and 'little draggings.' Then we get to the gyroscope, same principle working there.

A meteor. (Figure 12.1) Illustration here of a meteor traveling through space. Its dynamic energy is conserved by all the little Space Particulates pulling and dragging on it, the more mass it has, the greater number of hard matter atoms passing through the greater number of space particulates that the meteor's volume will occupy, or pass through, and therefore more little pullings and draggings. A potent force.

The Real Universe.

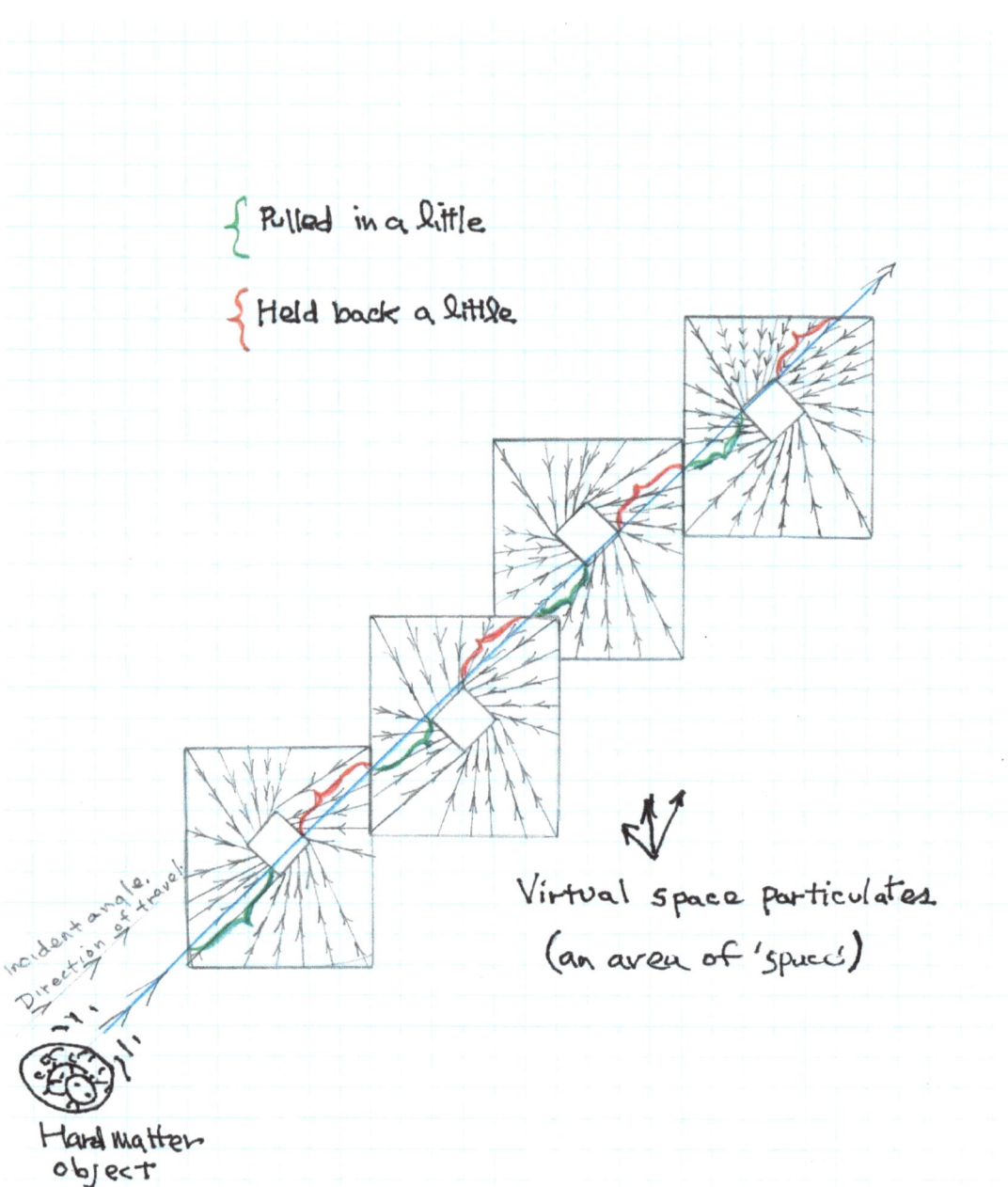

Pulled in a little

Held back a little

Incident angle.
Direction of travel.

Hard matter object

Virtual space particulates
(an area of 'space')

Figure 12.1

The Real Universe.

13. Light Speed and Newton's Cradle.

Like hard matter, light travels through space particles. Light is also affected by the structure inside a given space particle. As has been set forth, the filaments inside a space particle have potent effects on what is traveling through them, hard matter or light. At this point light traveling through space is being considered. It is theorized in PC that space particles have a structure. In the space particle, the filaments and nucleus are proposed to have a directionality influence. They either pull in or push out things passing through them.

Light, photons of light, or a beam of light is postulated to be transmitted by a method similar to a Newton's Cradle (Figure 13.1.A). This is a bit more involved than just energy impacting one ball and being transmitted through the elastic steel to the next ball. Individual space particles are comparable to the steel Newton balls in the PC theory of light transmission and speed.

As seen in the illustration (13.1.B), light or light energy of a photon of light impacting and entering a space particle (or Newton Cradle ball) has the particle's internal structure to traverse. The energy of the photon impacts the particle and is transmitted through the particle in a manner not unlike lightning. The photon energy seeks the easiest path or path that is most conducive to the traversing or conduction through the particle. The easiest path is along the filaments and core surface as they are 'denser' spatial fluid. This could be comparable to current flowing along inside copper wire. The angle of incidence of the traveling photon is the same exiting the space particle as it was entering the particle. This is illustrated in figure 13.1. Though the photon is 'traveling' or being transmitted in the same line of travel, it is now parallel to its original angle of incidence. The photon has been 'moved over' a little by the amount of the detour it took. For the photon to travel in a straight line (Which it does) it needs to be 'deflected back'. This happens in subsequent space particles. So the travel of light consists of a vast

The Real Universe.

number of detours and/or re-routing. The light keeps traveling straight by detouring in complementary angles.

Light passing through a space particle at a given angle of incidence travels along filaments and the surface of the space particle nucleus. Flowing along directionally influenced filaments or being pulled in a little and being held back a little as happens with hard matter is not postulated to be as influential as the 'detour' the photon takes through the space particle.

It can be seen geometrically that the photon, by following the surface of the nucleus of the space particle has taken a detour. The photon angles off and then angles back to its course, then does the opposite in a subsequent spatial particle to continue moving in a straight line. To simplify the example, only 'deflection of conductivity' caused by the core of the space particle is considered here. There well may be and PC postulates that there is, 'deflection of conductivity' caused by the filaments of the spatial particle that will not be dealt with at this writing for simplicities sake.

So the photon transmits its energy or travels across or through space or through each space particle in the same manner as energy travels through a Newton's Cradle. The contiguous spatial particles being compared to the balls in a Newton's Cradle. This can be seen to happen at a great speed.

The detouring off line of direction and back to original direction along the surface of the space particle nucleus as the photon goes through each particle takes time. This time of detour and back, traveling through each particle then, is a powerful governor on the speed that the photon travels and therefore the speed of light.

It can be seen that if the nucleus of the spatial particulate where smaller, the speed of light would increase. Further, as the nucleus decreases in size, the speed of light would increase, perhaps with no nucleus the speed of light would be or approach infinity.

The Real Universe.

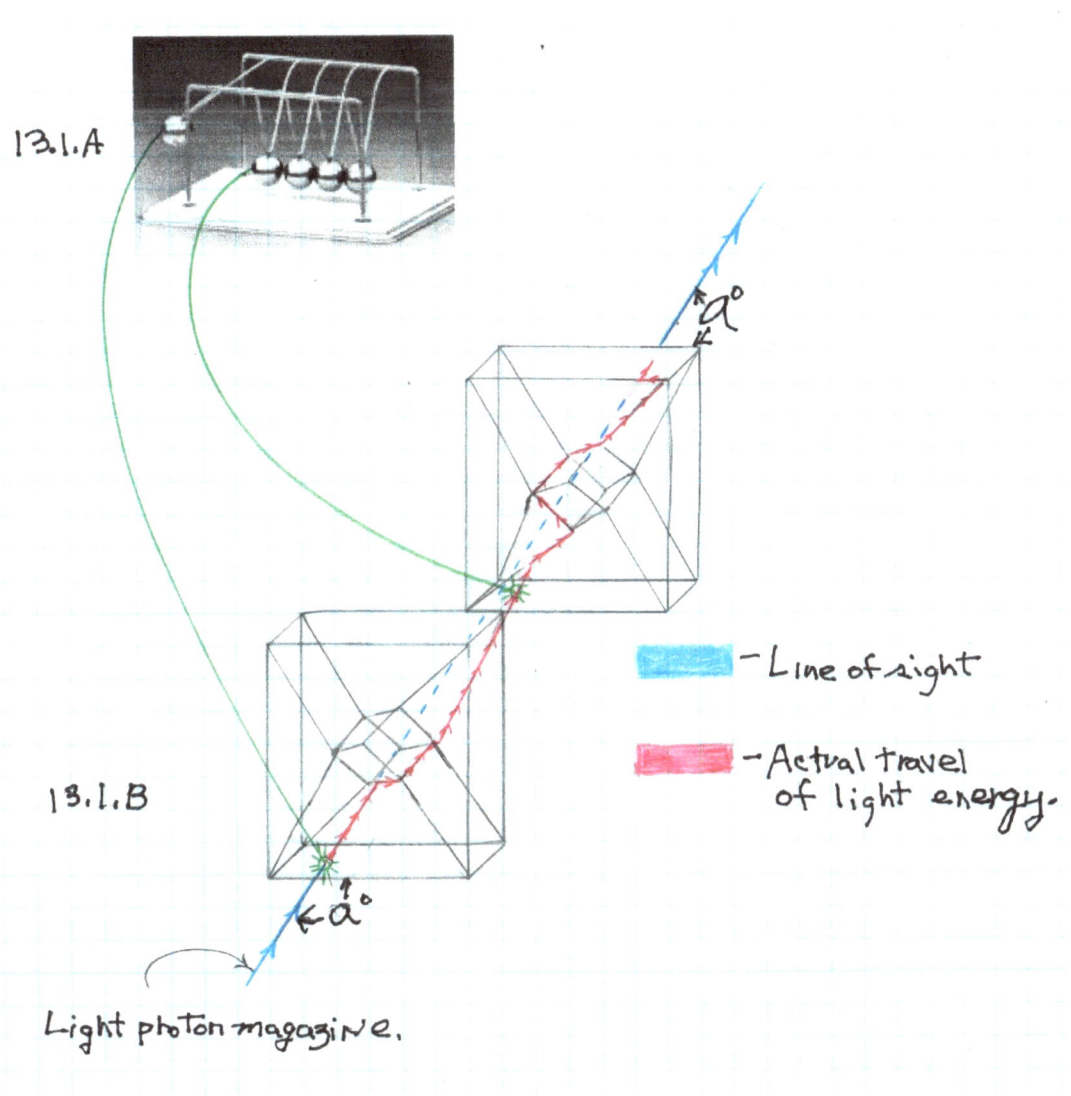

13.1.A

13.1.B

— Line of sight

— Actual travel of light energy.

Light photon magazine.

Figure 13.1

14. The Spectrum. The house of mirrors. More Newton's Cradle.

The (A) spectrum is created as the result of passing through or traversing a (or multiple) Spatial Particulates. A photon of light contains a lot of information, much more than is transmitted through spatial media. With the medium to heavy elements there is a profuse amount of information released in the elements vibrations or even it's existence. The information that does make it across the spatial medium is focused or deflected by the nucleus of the spatial particulate. Most information would be blurred if it merely traveled on it's own across some theorized void. Instead The Particulate acts as a deconstructing/re-focusing lens if you will. The spectrum information that we have is not contained entirely in the photon.

As seen in the illustration the energy/information of the photon for any given element passes through a particulate or a multiple of particulates acting as one. From any given element there are a 'set' of wavelengths particular to itself. They can be seen encountering PC's spatial particle. An absolute spectra for that element is created by the different lengths or energy segments of that element's vibrations interacting and/or being deflected and/or their plane of travel being 'shifted' by the internal structure of the spatial nucleus.
So the photon of energy is redirected entirely, or shifted out of its plane of travel (14.1 a.). There are shadows or diffusion of some lines of energy. This is why a prism 'reconstructs' (If you will) the spectrum. Some energy levels are deflected entirely (14.1 b.), this is most likely why a diffusion haze is seen.

Again, interesting fodder for crystals.

The Real Universe.

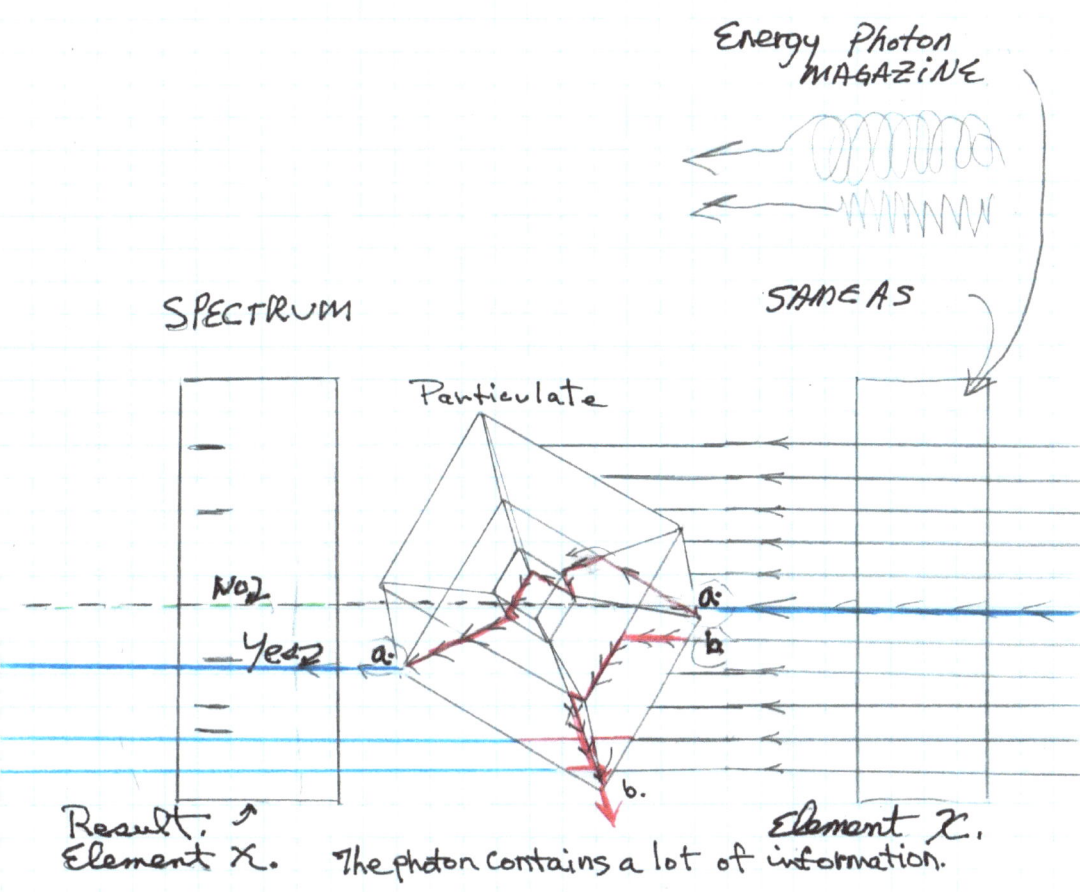

Energy Photon
MAGAZINE

SAME AS

SPECTRUM

Particulate

No2

Yes2

a.

a.

b

b.

Result: ↗
Element X.

Element X.

The photon contains a lot of information.

The spectrum is not entirely in the photon.

(a.) is shifted
(b.) is deflected entirely. Proly where
    diffusing is started.

- Space particulate is prism or
  house of mirrors.

Figure 14.1

The Real Universe.

15. The Atom.

PC will consider only the hydrogen atom for simplicity of postulation. In PC the atom is cubic since the cubic spatial particulate that was used as a mold to condense space and create hard matter (The proton and neutron) is cubic or polygonal. The atom (Here the hydrogen atom) is most basically a cube, atoms are not spherical they are cubical. The cubic atoms are contiguous. They touch on all sides. PC regards the orbital model of an atom as antiquated . Distance between atoms being held off or apart from each other by supposed electrical fields may work for say, chemical engineering as a model, however it is not the case that atoms are discontinuous and stand off from each other because electrical fields hold them apart.

In the antiquated orbital atomic model atoms not contiguous with each other would have a void in between them, this is an inconsistency with PC and any real physics that there would be any void at all, anywhere.

The series of illustrations (Figures 15.1) reiterates nucleosynthesis to the solid state hydrogen atom. The Particulate is shown. The neutron and electron are illustrated. The neutron having had an even number of 'moldings' giving it a 'right handed' bias of nucleus and filaments and the proton having had an odd number of 'moldings' giving it a 'left handed' bias. The proton and neutron are both quite unstable and oppositely threaded. The neutron and proton then 'screw together' being threaded into each other to create the hydrogen atom.

It is postulated that the filaments are what is believed to be electrons. Heavier elements are created in a furtherance of synthesis with hydrogen as the basic raw material in the crucible cores of stars. Being that atoms are created by very large amounts of energy fusing them or fusing into them, when an atom is split or fused a lot of that energy is released.

Given PC's polygon steady state nature of the atom or nuclear structure, the toroid is submitted for consideration.

The Real Universe.

With plasma confinement by using toroidal magnetic fields, or Tokomak. In the (15.1) it can be seen that a toroidal field is in a way similar to atomic structure. Holding plasma in a Tokomak is actually close to holding it in the shape of an atom. This toroid seems to be reproduced in nature. That is why the Tokomak seems promising.

The earths electrical field is submitted as an example, it appears to be toroidal in configuration. PC holds that the earths magnetic field is really cubical, reproduced on an immense scale. The cubic field is very malleable, dynamic and of course, flexes and bows from internal and external influences, but it is held that the true shape of the earths electrical field is a cube.

Also, crystals or crystalline structure seem to be another example of the macro mirroring the micro. Instead of theorizing of how the molecules of a crystal structure are interpolated into cubes, understand that it is the cubes that are the molecule or atom

The beehive is telling of polygon solid state nuclear structure. Nature seems to know that there is no void or waste.

The object at the north pole of Jupiter is also submitted for consideration. How in nature does any polygon exist unless the nuclear building blocks are polygonal themselves?

The Real Universe.

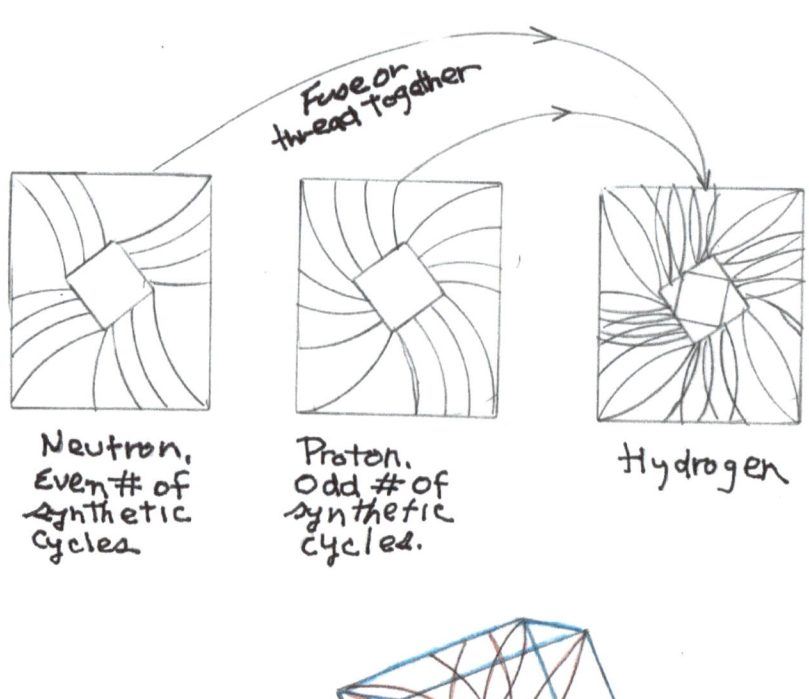

Fuse or thread together

Neutron.
Even # of
synthetic
cycles

Proton.
Odd # of
synthetic
cycles.

Hydrogen

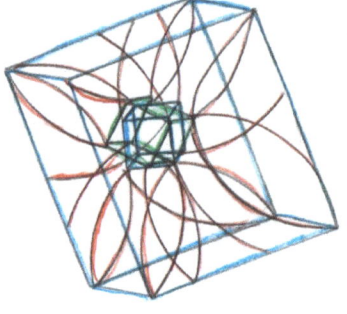

Hydrogen Atom

Figure 15.1

The Real Universe.

16. Pi. π

So then, its on to a cubic and finite light horizon bubble (LHB) universe. Math explains events that happen in the physical world. Math explains the physical structure of our universe. There can be no deception or mistake if we pay attention to the math. The PC model of our universe (Our light horizon bubble) begins with <u>The</u> energy event or a big bang. The PC big bang is rather like a Hoberman Sphere expanding outward. More like a Hoberman Sphere than a real theoretical 'sphere' expanding because, in PC there is no such thing as a 'sphere'. All space particulates and hard matter are cubes or polygons.

Mathematics teaches us that circles or spheres are not real, but only theorized objects. π is the ultimate authoritative informer that a light horizon bubble, or star or planet are not spheres, that there are no true hard matter or spatial matter spheres. It decrees by mathematical law that the ratio of a circles circumference to its diameter will only be rational when the segments Archimedes inscribed inside his circle equal the length of a side of a particulate. The length of a side of a space particulate is small, but not infinitely small. If π is carried out far enough, it will end, giving the length of a side of a spatial particulate. So if any circular objects surface is looked at closely enough it will be seen to have a jagged, saw-toothed edge. These are the corners of the space particulates, or the corners of the atoms an object is made of. That the surface of a sphere is smooth is only theoretical. (Figures 16.1)

The Real Universe.

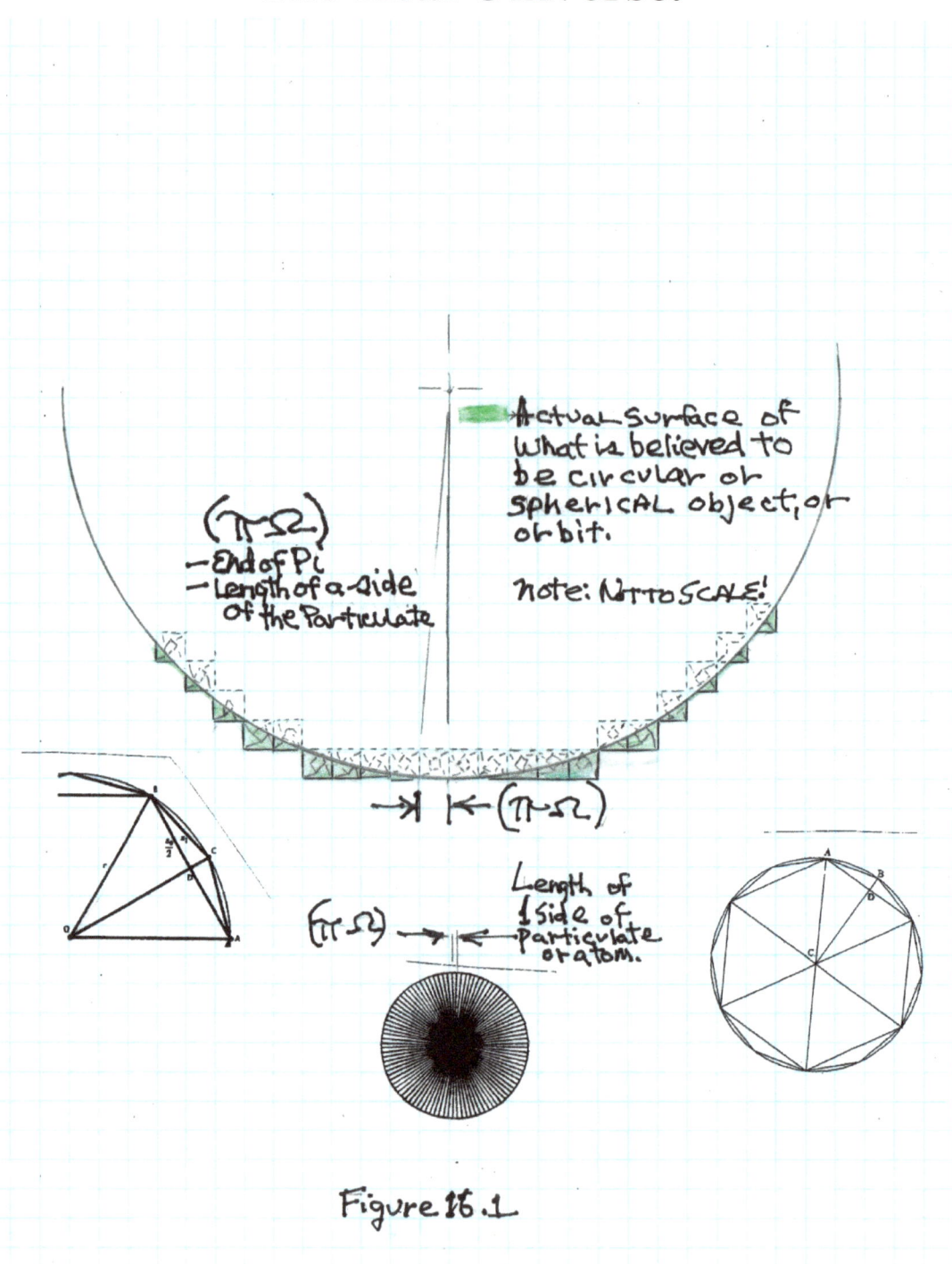

$(\pi$-$\Omega)$
— End of Pi
— Length of a side
  of the Particulate

Actual surface of
what is believed to
be circular or
spherical object, or
orbit.

note: NOT TO SCALE!

$\rightarrow\!\!\models\!\!\longleftarrow (\pi$-$\Omega)$

$(\pi$-$\Omega) \rightarrow\!\!\models\!\!\longleftarrow$ Length of
1 side of
Particulate
or atom.

Figure 16.1

The Real Universe.
17. Time, DNA and Time Travel.

PC defines time as an 'Exact Movement.'

Time starts when there is hard matter and spatial particles, and the hard matter begins to move. The spatial particles are in a fixed three dimensional lattice or fabric throughout our light horizon bubble. (Figures 17.1) Hard matter or matter as it is now understood 'moves' through space. At the same time, space or each space particle is so minute that it passes through the hard matter. Like a microscopically thin needle will pass through a person and won't be felt. It might be assumed that spatial particles would move out of the way of the object passing through. This is not the case. Spatial particles have the ultimate moments of inertia, they may distort in the presence of energy because of their near infinite elasticity or malleability, but at the same time their smallness facilitates their passing through hard matter.

This being the case time began where the first energy event occurred at (0,0,0) in 'space.' This is an exactly locatable 'place' as the infinite lattice of space particles have not moved, the exact location of the big bang is calculable using space particles as the calibration unit along three axis, it is still there and will always be there.

Our 'moving through space' is by no means random in our LHB. We rotate and orbit the sun, <u>and</u> the sun is in an orbit presumably around the center of our galaxy. These are three fairly certain movements we are making through the lattice of spatial particulates. Our galaxy could be orbiting also, but that will be considered superfluous at this time.

If drawn (Figure 17.1) the motion appears similar to DNA helix. PC speculates here that DNA is in some way deeply rooted to our motion in space as time literally goes by. It is postulated that this motion is deeply rooted in DNA creation and such ongoing movement is a significant cause of aging in that the twisting or bias of DNA is potently affected by intertwining or twisting with the bias angle of the spatial particulate. The interaction could cause the same action on DNA as a

The Real Universe.

flag flapping in the wind, essentially tattering DNA or virtually wearing or aging DNA and us.

Time travel is <u>not</u> straight forward in PC, as matter of fact traveling back in time is a matter of traveling the reverse of the way that one has been going. If one where to travel in reverse, of the earths rotation, the earths orbit and the suns orbit all at the same time it would be equivalent to going back in time. No one would be there as the earth and everyone on it have kept on going 'forward.' The same with going forward in 'time'.

Therefore, if space particles where used as units of measurement it is possible to track any movement using three axis coordinates with increments that are absolutely stable in our universe. The movement would have a (0,0,0) starting point and have an absolutely unchanging frame of reference to measure it with.

Pouliotion Physics is absolute not relative. It is specific not general, thereby being a, 'Theory of Absolute Specificity.' All matter, motion and time have their behavior accounted for. There is NO uncertainty, therefore an, 'Uncertainty Principle' is what is ridiculous, not Einstein's unfortunate phrasing of a very correct observation. You can apologize now guys.

The Real Universe.

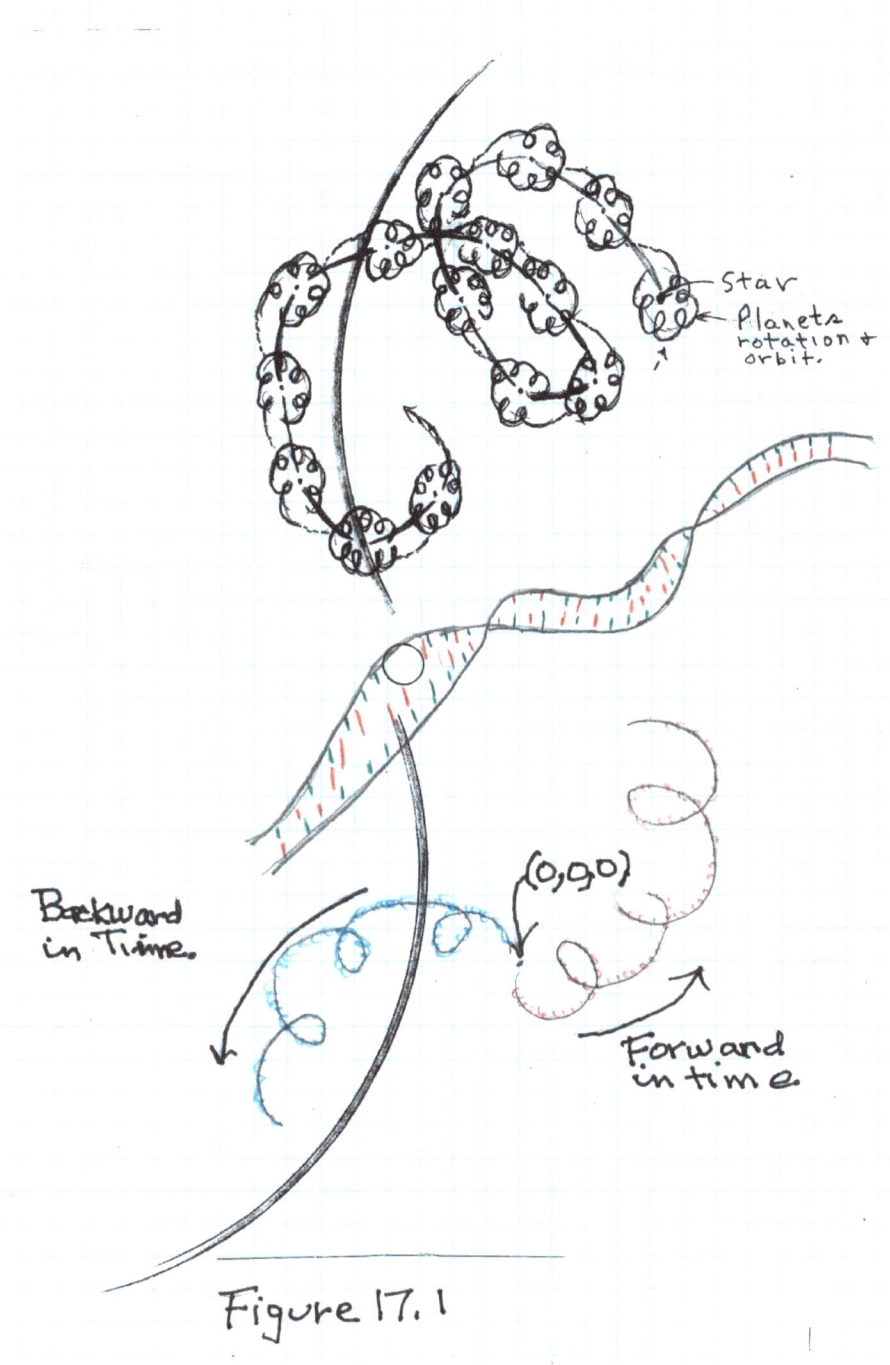

Figure 17.1

The Real Universe.

18. Free bonus chapter!

It is stated here that PC affords a deeper understanding of physics equivalent to medicine acquiring an understanding of bacteria and viruses. To mention a few things that new understanding facilitates: Fractals, Kirlian photography, Hurricane and ocean vortexes. Electricity can be seen to have more to it than is presently known or theorized. It sheds new light on heat transmission from the sun, and lightning . Solar panels will improve greatly.

It is postulated that what we believe to be our universe is actually the real, 'Island Universe' Hubble thought at first he was seeing. There are other Light Horizon Bubbles created by other energy events, other big bangs. It makes bracing sense in an infinity of space that there would be more than one energy event in limitless time. Further, if there is more than one big bang in space, there are infinite big bangs. PC's author wonders if some cosmic observations have seen out of our Light Horizon Bubble and into an adjacent one? (Figure 18.1)

The Real Universe.

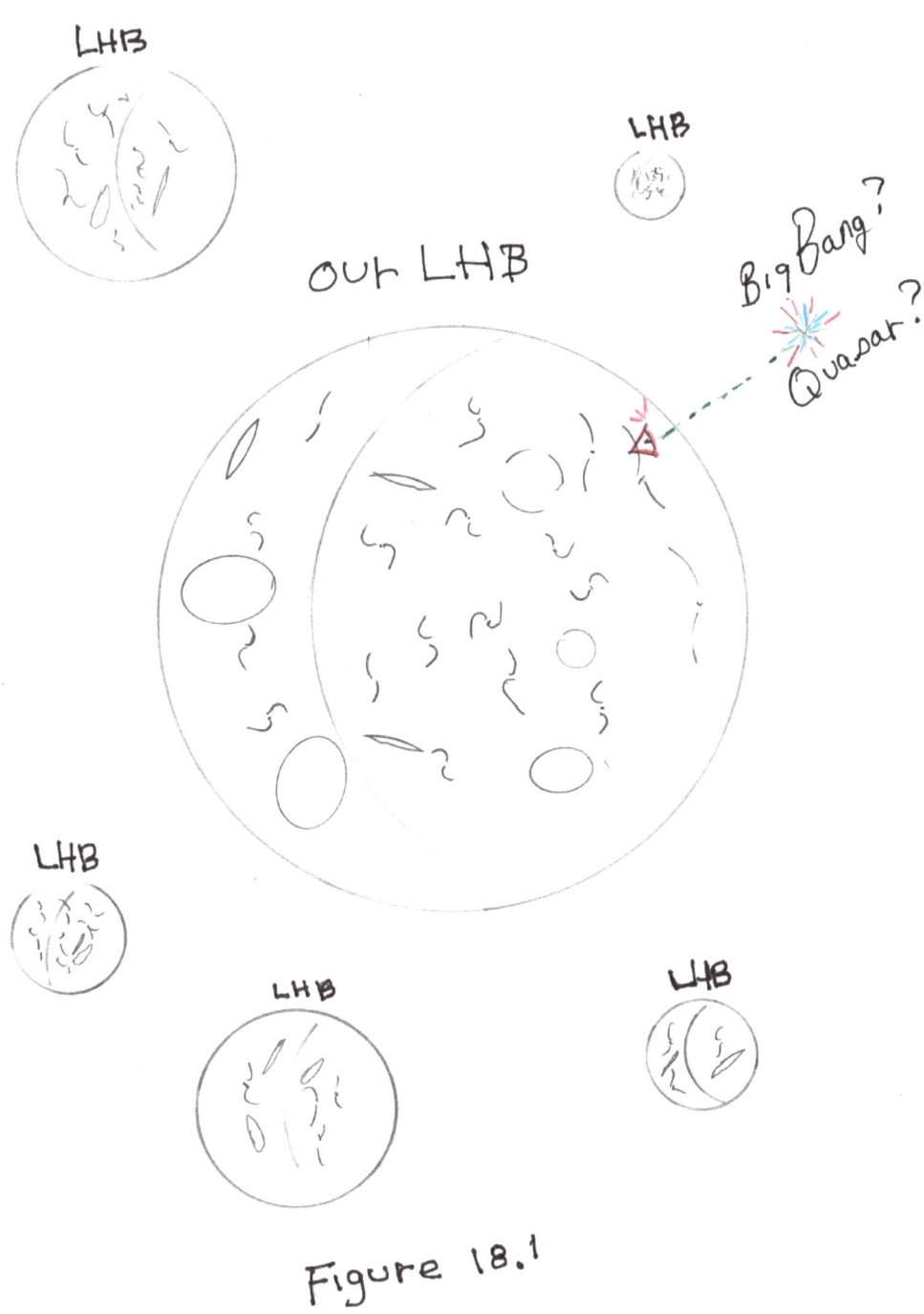

Figure 18.1

The Real Universe.

19. Communication to other Earths.

Hold a radio up to a running motor or generator and you'll hear static, the radio will be mostly useless. The metaphor is using a radio telescope to listen to the universe and having at least three big generators running. The earths rotation, the earths orbit and the suns orbit are the generators I speak of. They all create static on our end. Presumably anyone on the other end would have the three same generators interfering with their reception and transmission. To hear someone on the other end, both parties would have to null out or filter at least three main interferers each. (Figure 9.1)

Communication across the LHB?

The Real Universe.

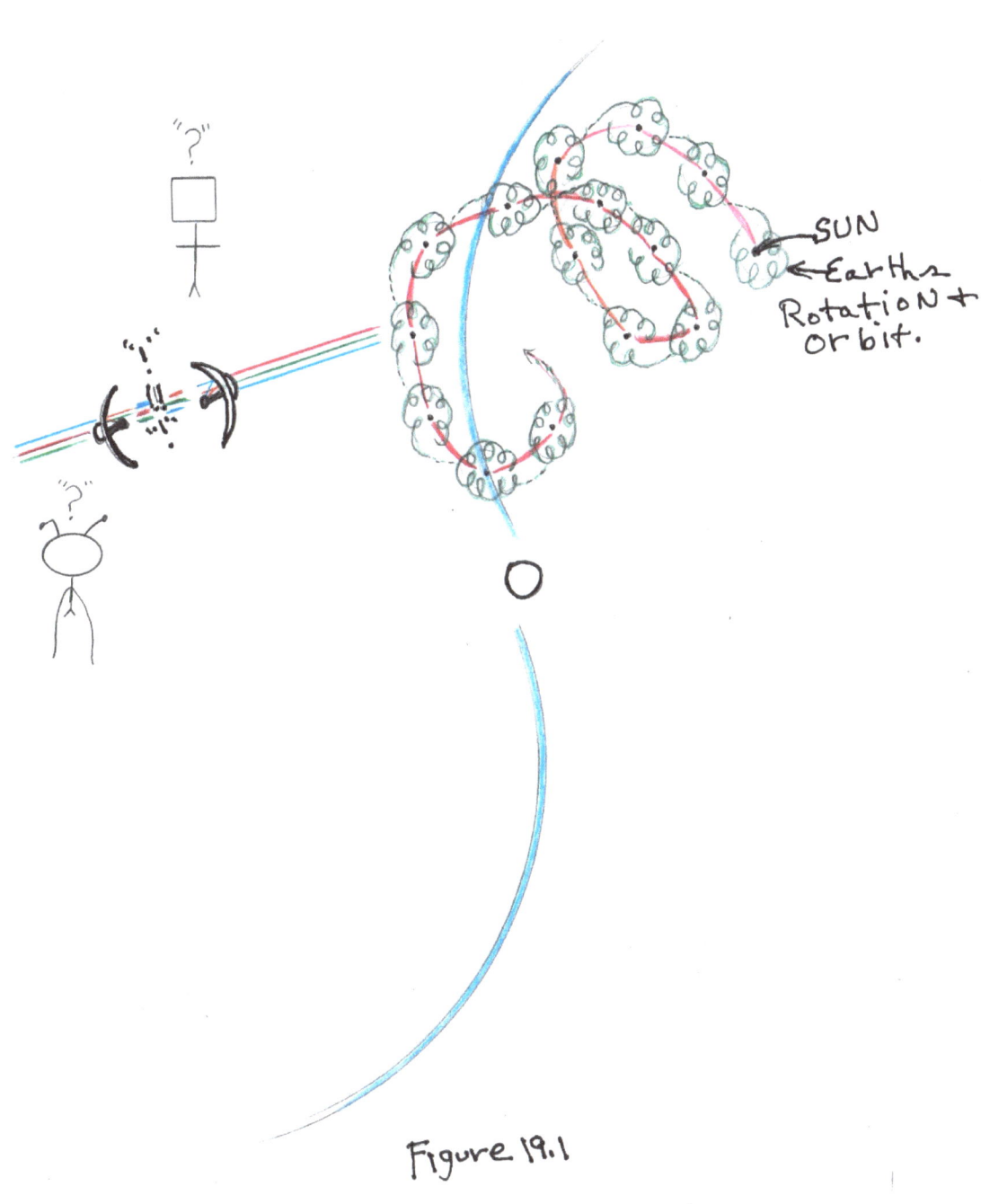

SUN
Earths
Rotation +
Orbit.

Figure 19.1

The Real Universe.

20. How was the shape of The Particulate deduced?

To understand the micro is to know the macro. The macro is telling of the micro. With back and forth observing the micro and the macro, one tells of the other and vice versa. Macro actually displays what is going on at the smallest levels.

The Particulate shape deduction began with the impossibility of a void. With any orbital or mythical circular atom there would be a void, so the polygon. A cube within a cube here for simplicity. Present state of the art torroidal considerations gave credence and strength to the cubic model with the knowledge that there is no void, anywhere.

Again with the macro an indicator of the micro, the nucleus replicating the mirror image of itself repeatedly, some key observations where the finishing confirmations on this theory.
Consider this super nova: (Acknowledgment to Hubble)
(Figure 20).

The Real Universe.

Figure 20.

hubble-08-cats-eye-nebula-1080v.adapt.710.1

The Real Universe.

You can see here (Figure 20) The Particulate formation replicated in immense struts in multiple basic cube within a cube configurations. This depth charge implosion nucleosythesis is ongoing. It is a smaller version of the big bang, if you'll excuse me for saying so. Nevertheless it is creating 'Hard Matter' and also one of many creating gravity by adding more ice (Hard Matter) to the water (Spatial Particulate/Fabric/Fluid). Vast ribbons of hydrogen are being created, spiraled out.

Particulates are being synthesized repeatedly, as long as there is enough energy. Results of an even number of compactions are neutrons, an odd number of compactions, protons. The protons and neutrons 'thread' together forming hydrogen.

I've tried to outline the smaller block units of hydrogen formation in Figure 20.1

In Figure 20.2A the ribbons of hydrogen are visible going out at millions of miles per hour from the side view. It is a good cut away of what is happening in a sphere. This super nova seems to have had multiple implosions judging by the several concentric rings radiating out from about six bulging sections in the center that are still in quite a bit of chaos.

The areas outlined in green (Figure 20.2.B) seem to be areas where the odd/even synthesis has not yet reached reverse meta-stability and neutrons and protons are being created and threading into each other to create hydrogen.

The Real Universe.

Figure 20.1

The Real Universe.

The Real Universe.

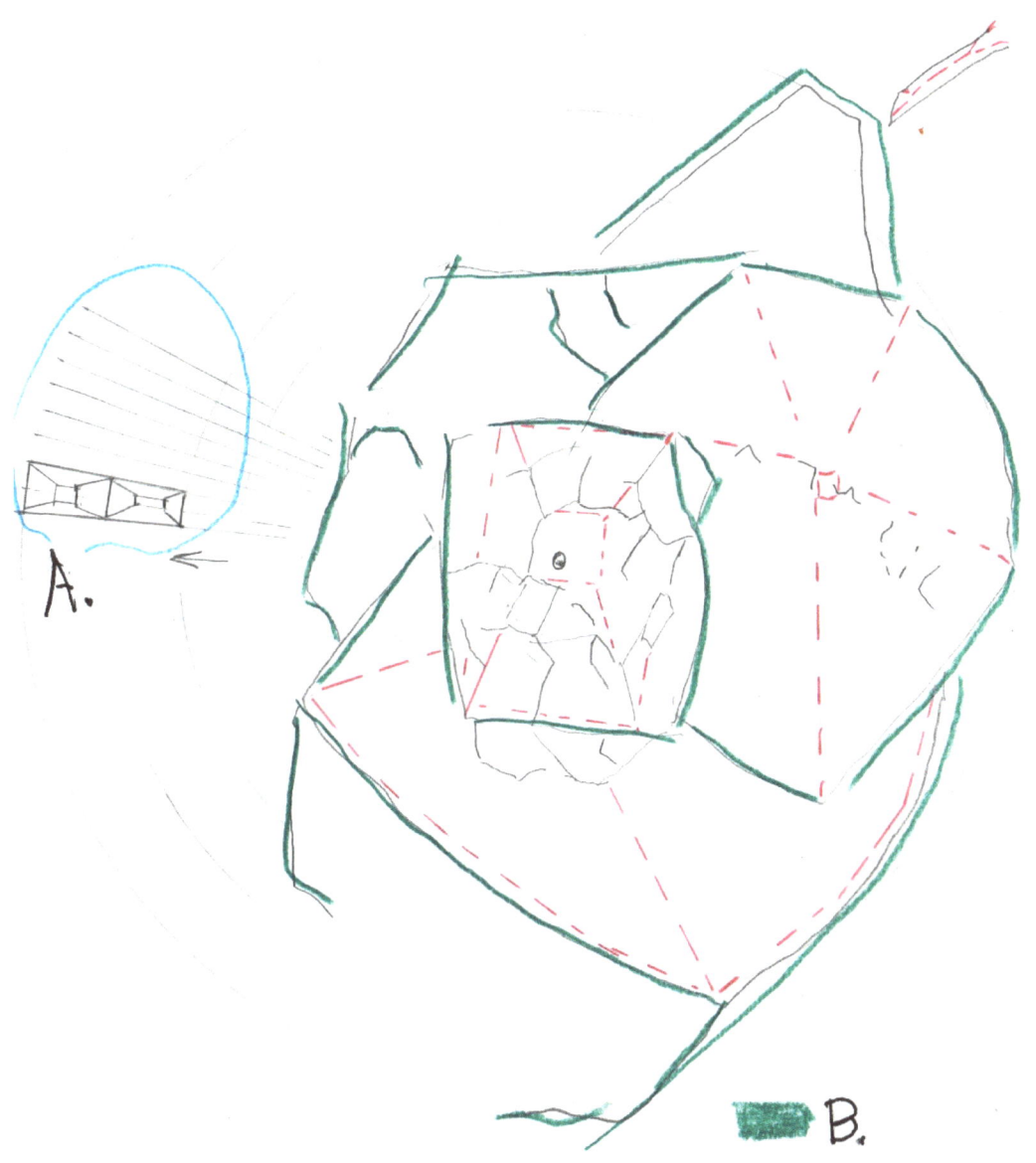

A.

B.

Figure 20.2

The Real Universe.

Epilogue.

This hypothesis is just a start. It is obvious there is no math here. Though I went on to complete all four courses in calculus, my training is in machining, my B.S. in Business Administration, I've also trained in diesel mechanics and vocal music (Mostly the latter). This is only an introduction given to me to introduce to young people that have been looking for it and will take it and run with it.

The theory is simplified. There is no claim that all of this is correct. The foundation premise is most firmly held as correct, but it is a start.

I have envisioned as the result of the founding of insight of the principles discussed herein, the birth of clean, endless energy. A new economic base for the foreseeable future. Medical knowledge of a truly advanced race. Interstellar propulsion undreamed of. A clean healthy earth. A chance to listen in on cosmic conversations that must be out there.

With only a small part of the above, I envision more time for our children, and more quality time for us with them.

I am a crooner with a keen and penetrating mind. I am a big hearted, very spiritual man who must have volunteered to go through refiners fire many times to be a vessel worthy to bring this knowledge to this time and place.

My thanks to all whose shoulders I've stood on. I want to acknowledge the man who taught me how to live without alcohol.

I want to acknowledge my son with his mind boggling courage.

Lastly, for emphasis I thank my angel wife. I have no words.

Truly my best wishes to all.

Russ Pouliot.

The Real Universe.

" - - - We will go down, for there is space there, and we will take of these materials, and we will make an earth whereon these may dwell; - - -"

Abraham 3:24.

www.ingramcontent.com/pod-product-compliance
Lightning Source LLC
Chambersburg PA
CBHW050751180526
45159CB00003B/1426